308410
$10.95
5-13-98

Y0-AEE-576

DATE DUE		

JUVENILE
591.57   Stone, Lynn M.
STO   Talons, beaks,
          and jaws

Snyder County Library
Selinsgrove PA

TALONS, BEAKS, AND JAWS

ANIMAL WEAPONS

Lynn M. Stone

The Rourke Press, Inc.
Vero Beach, Florida 32964

© 1996 The Rourke Press, Inc.

All rights reserved. No part of this book may be reproduced or utilized in any form or by any means, electronic or mechanical including photocopying, recording or by any information storage and retrieval system without permission in writing from the publisher.

PHOTO CREDITS
© Mary Ann McDonald: title page; © James P. Rowan: p. 21; all other photos © Lynn M. Stone

Library of Congress Cataloging-in-Publication Data

Stone, Lynn M.
   Talons, beaks, and jaws / Lynn M. Stone.
     p. cm. — (Animal Weapons)
   Includes index
   Summary: Describes how animals without teeth can defend themselves.
   ISBN 1-57103-166-9
   1. Claws—Juvenile literature. 2. Bill (Anatomy)—Juvenile literature. 3. Jaws—Juvenile literature.
[1. Claws. 2. Bill (Anatomy). 3. Jaws.]
I. Title II. Series. Stone, Lynn M. Animal weapons.
QL942.S76 1996
591.57—dc20   96-8999
                                                    CIP
                                                    AC

Printed in the USA

SNYDER COUNTY LIBRARY
1 N HIGH ST
SELINSGROVE PA 17870-1599

TABLE OF CONTENTS

Talons, Beaks, and Jaws	5
Bird Beaks	6
Swimming with Spears	9
Walking with Spears	11
Hunting with a Net	14
Talons	16
Birds of Prey	19
Toothless Jaws	20
Turtle Jaws	22
Glossary	23
Index	24

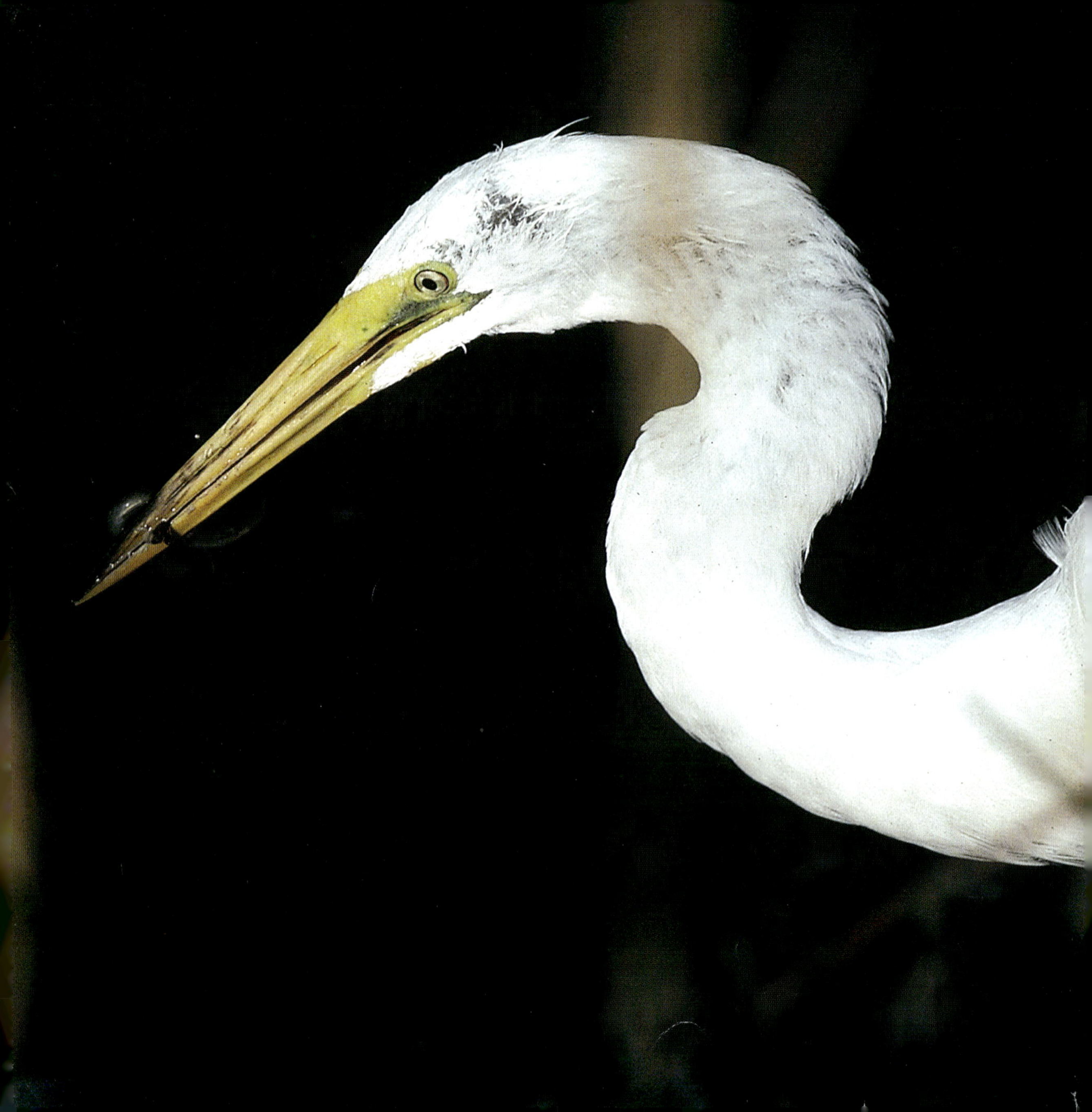

TALONS, BEAKS, AND JAWS

For animals without teeth, beaks, **talons** (TAL unz), and toothless jaws are important weapons. With these weapons, certain birds, insects, and turtles can be **predators** (PRED uh turz), or hunters, even without teeth.

Predators live on meat. They must kill other animals to survive.

Predators use their weapons mostly to kill. Weapons also help predators to defend themselves against enemies and their own kind.

Success! A great egret grabs an eel in its sharp beak

BIRD BEAKS

Birds have an amazing variety of beaks. Each type of beak is designed to catch a certain kind or size of **prey** (PRAY).

The beak of the cormorant, for example, has a hooked tip to help snag fish. The ibis has a long, down-curved bill. It neatly grabs shrimp or slips into a crab's burrow.

Cranes have long, fairly sharp bills. Cranes catch frogs, snakes, salamanders, and small creatures that live in mud and shallow water.

For a young sandhill crane learning to hunt, a ribbon snake is a slippery prize to swallow

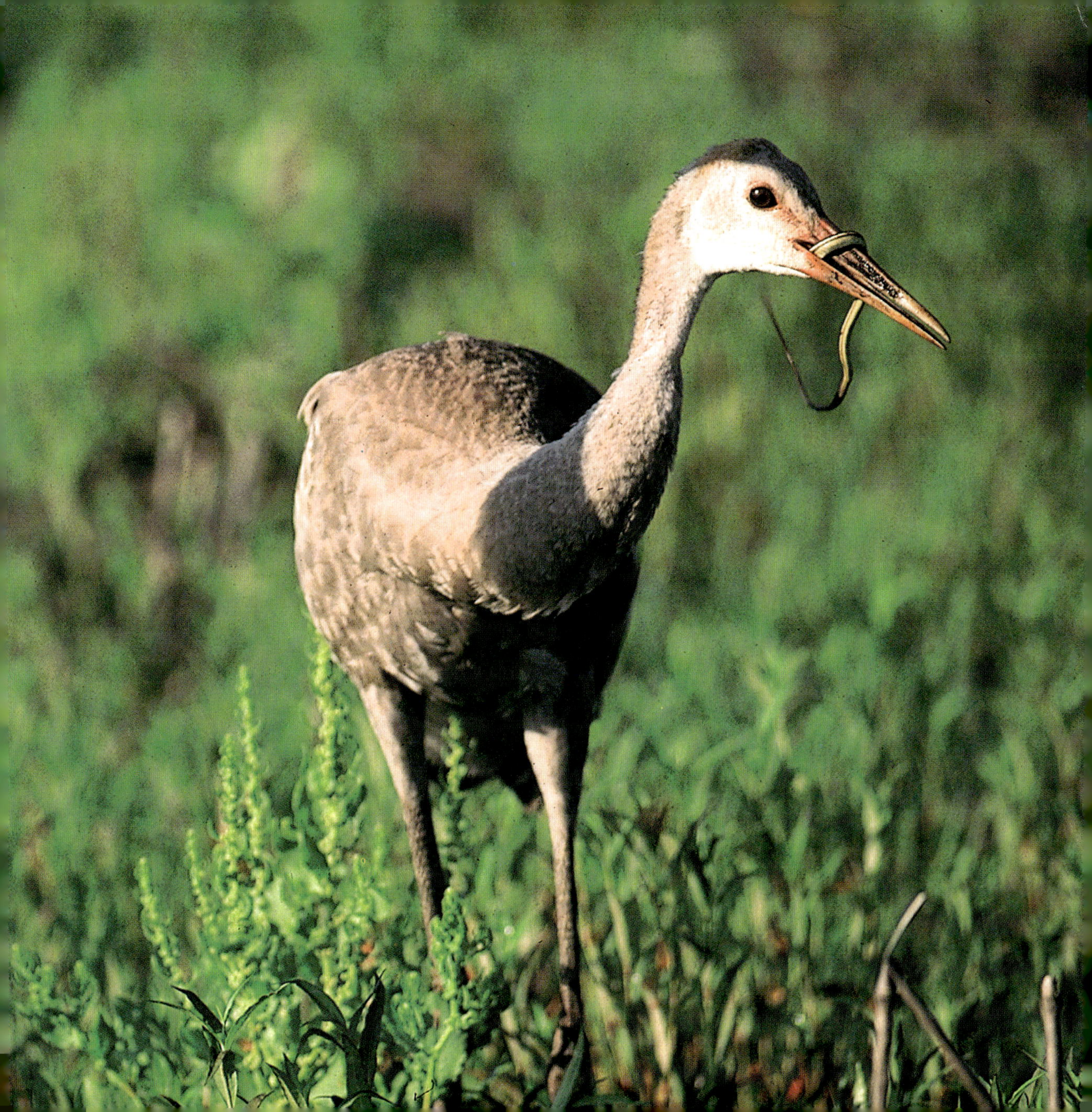

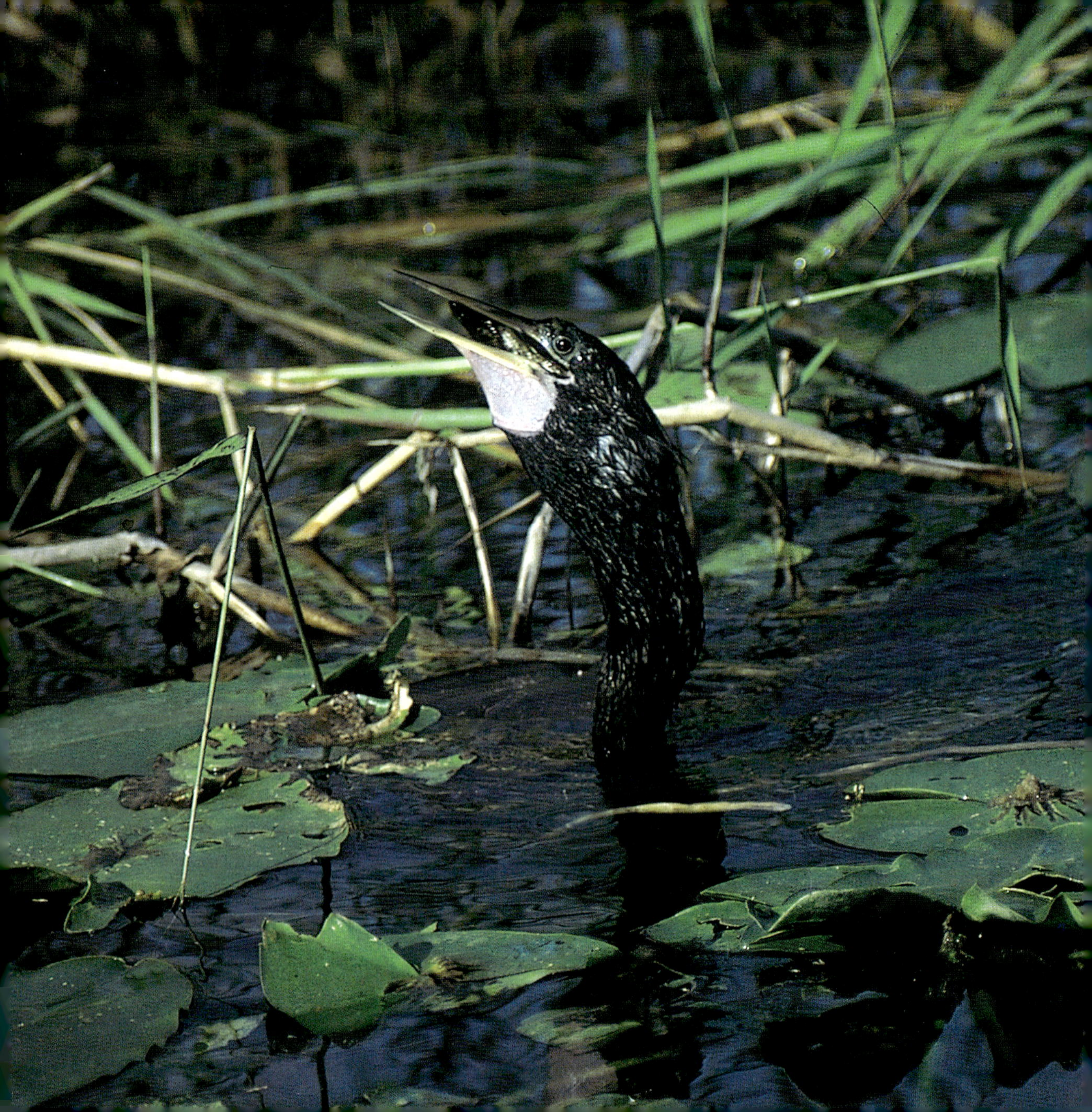

SWIMMING WITH SPEARS

Like the cormorant, the anhinga is an underwater hunter. It swims quickly by pumping its webbed feet.

At the end of the anhinga's long, snaky neck, its beak is a **lethal** (LEE thul), or deadly, spear to a fish.

The anhinga usually uses its bill to stab. It often brings its fish catch to the surface. The anhinga shakes the fish free, then grabs it head-first to be swallowed.

An anhinga pops up to snack on a sunfish in Florida's Everglades

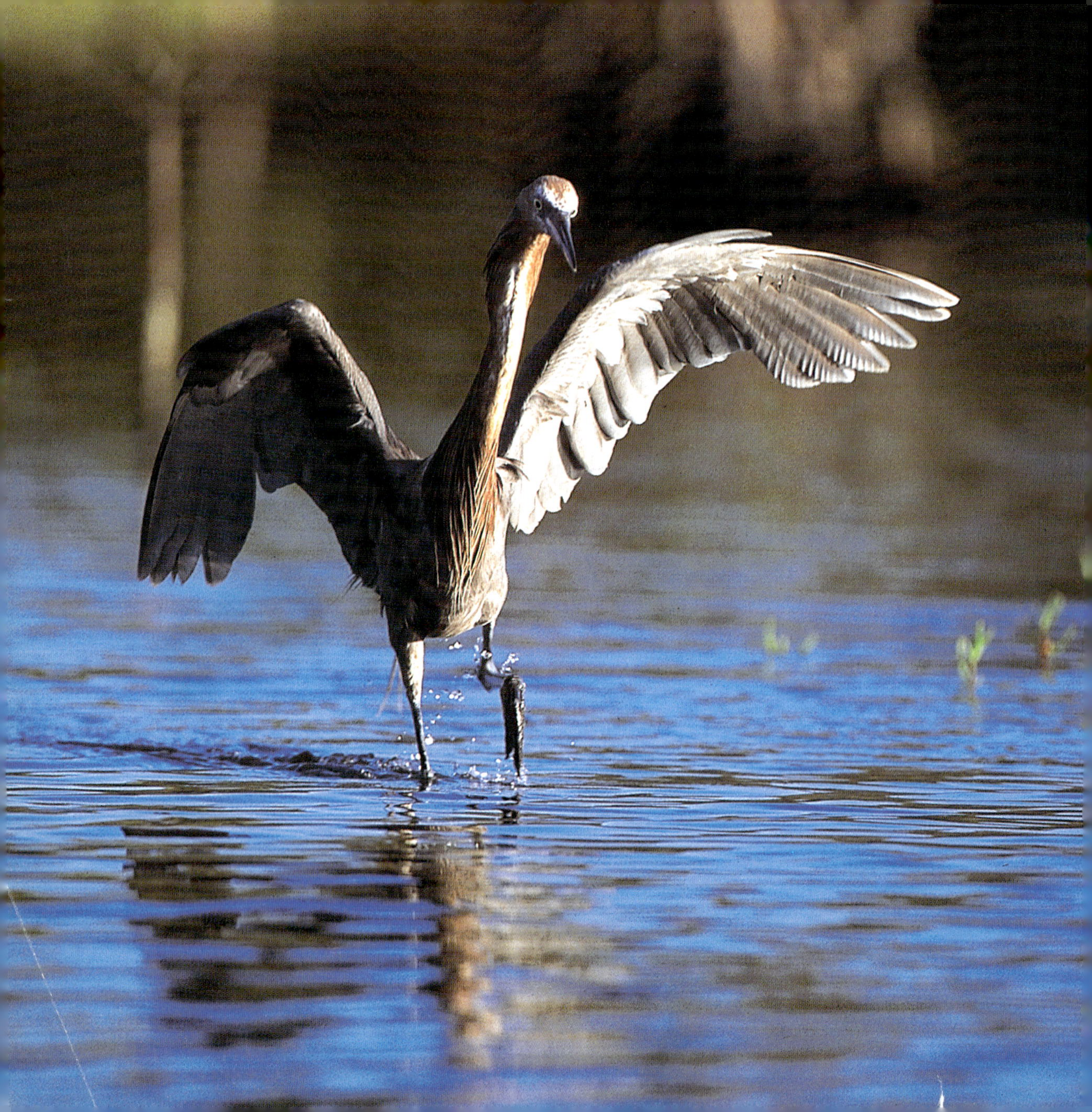

WALKING WITH SPEARS

Herons and **egrets** (EE grits) have beaks built for spear-fishing, too. Instead of diving, these long-legged birds wade in shallow water.

Most wading birds are very patient hunters. They may stand without moving for several minutes. They wait for a fish or frog to swim close. At the right moment, the bird thrusts its spearlike beak at the target.

The reddish egret is the odd duck of the family. It chases fish!

A reddish egret chases a school of small fish

A bald eagle finishes a high-speed attack by striking a fish at the ocean surface...

. . . and flying off with its trophy

HUNTING WITH A NET

One bird's lethal beak looks quite harmless, even comical. It is the pelican's pouch.

The pouch is made of thin, loose skin. It's attached to the pelican's lower jaw and neck.

The pouch stretches open, and the pelican uses it like a scoop or net. The pouch traps fish along with a "bucket" of water.

The pelican lifts its head to drain the water. Then it tosses the fish down its throat.

A brown pelican's pouch traps a luckless fish

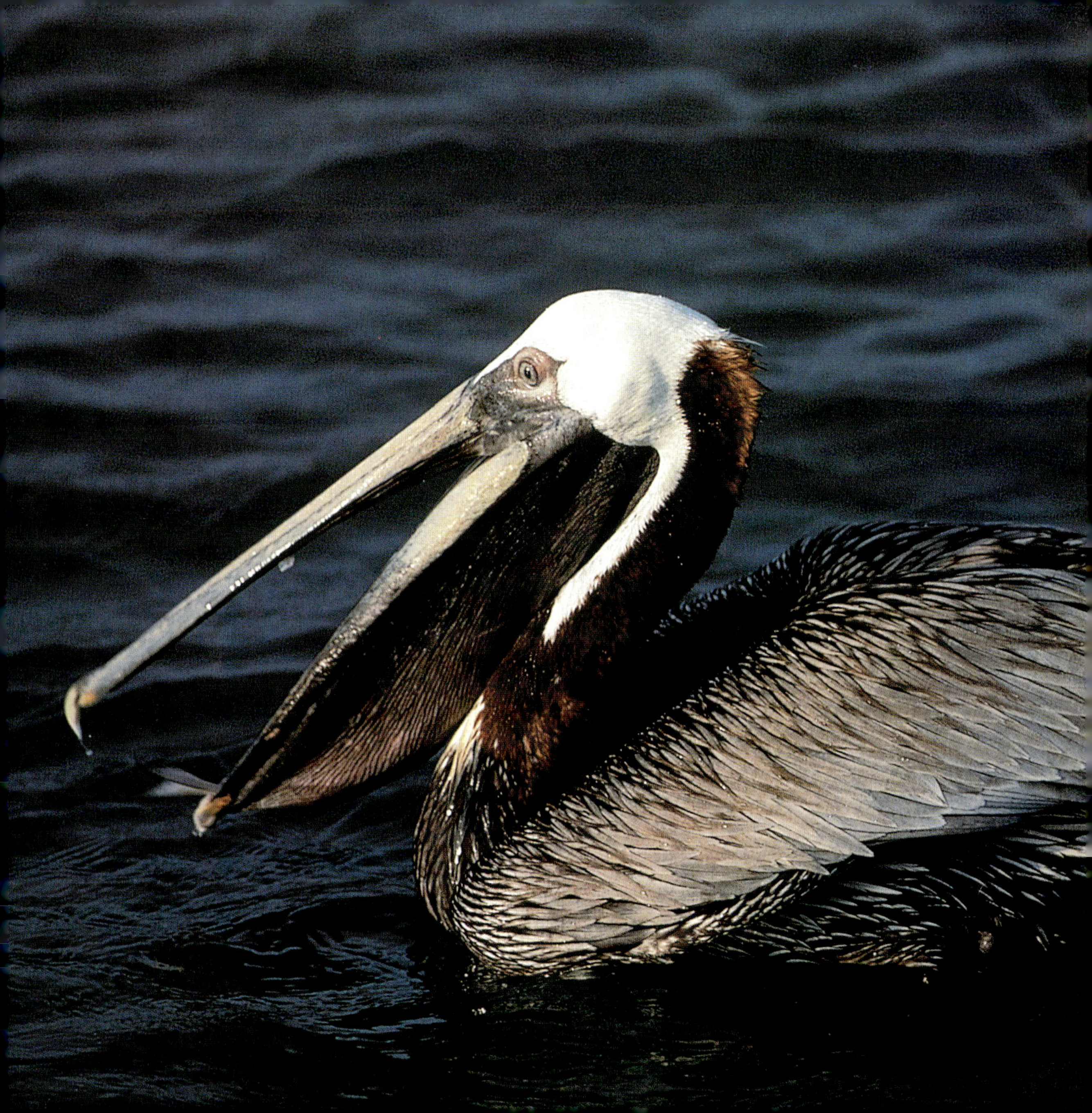

TALONS

Talons are the sharp claws on the toe tips of birds of prey. Birds of prey are the eagles, hawks, vultures, and owls.

Vulture claws are weak, but other birds of prey have truly lethal talons. Birds of prey kill a wide variety of animals, from insects and fish to snakes, ducks, monkeys, and rabbits.

A bird of prey uses its talons like powerful, hooked hands. It strikes an animal with open talons, then instantly grips its prey.

Talons give hawks, eagles, and owls a mighty grip and instant killing power

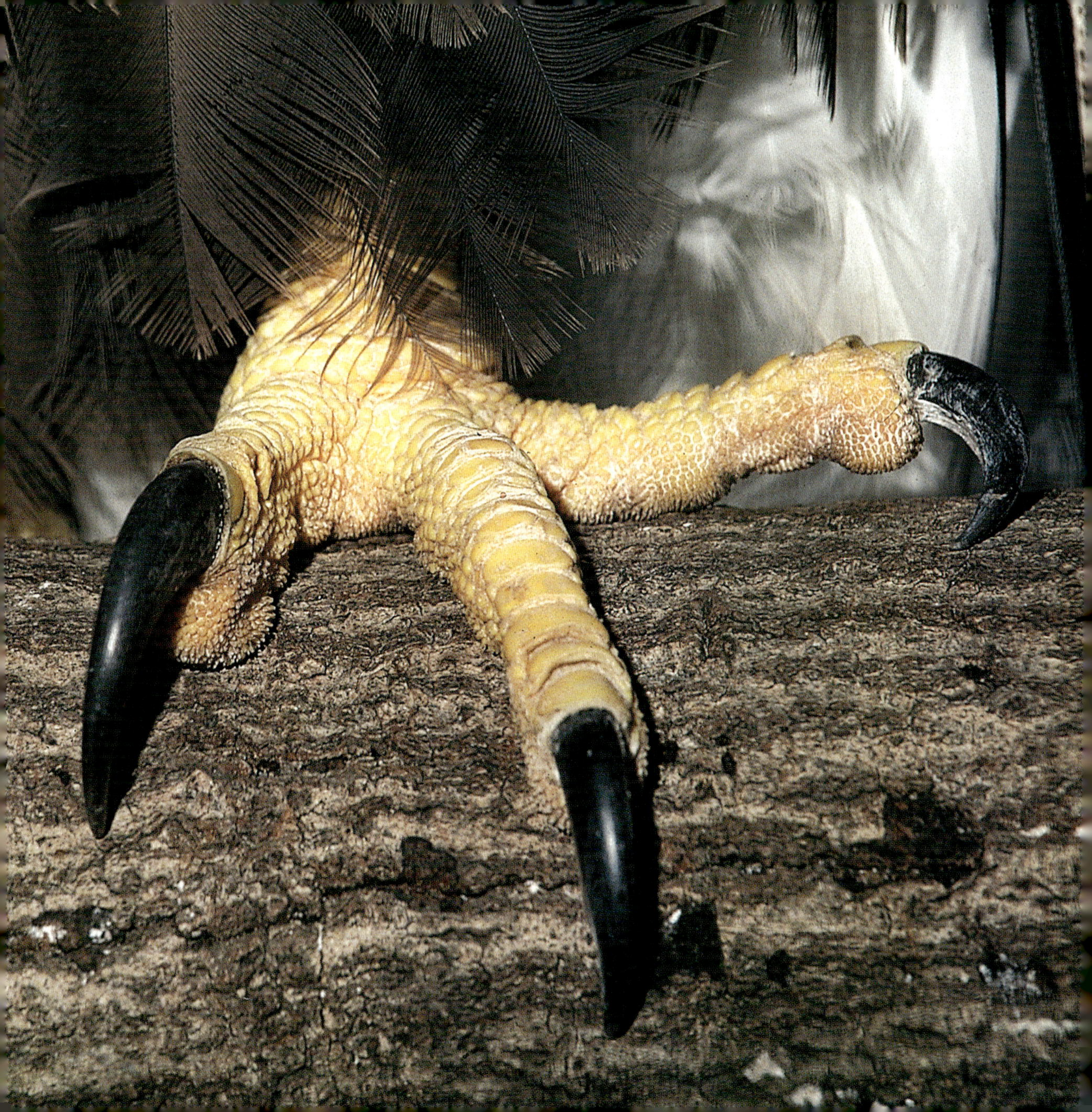

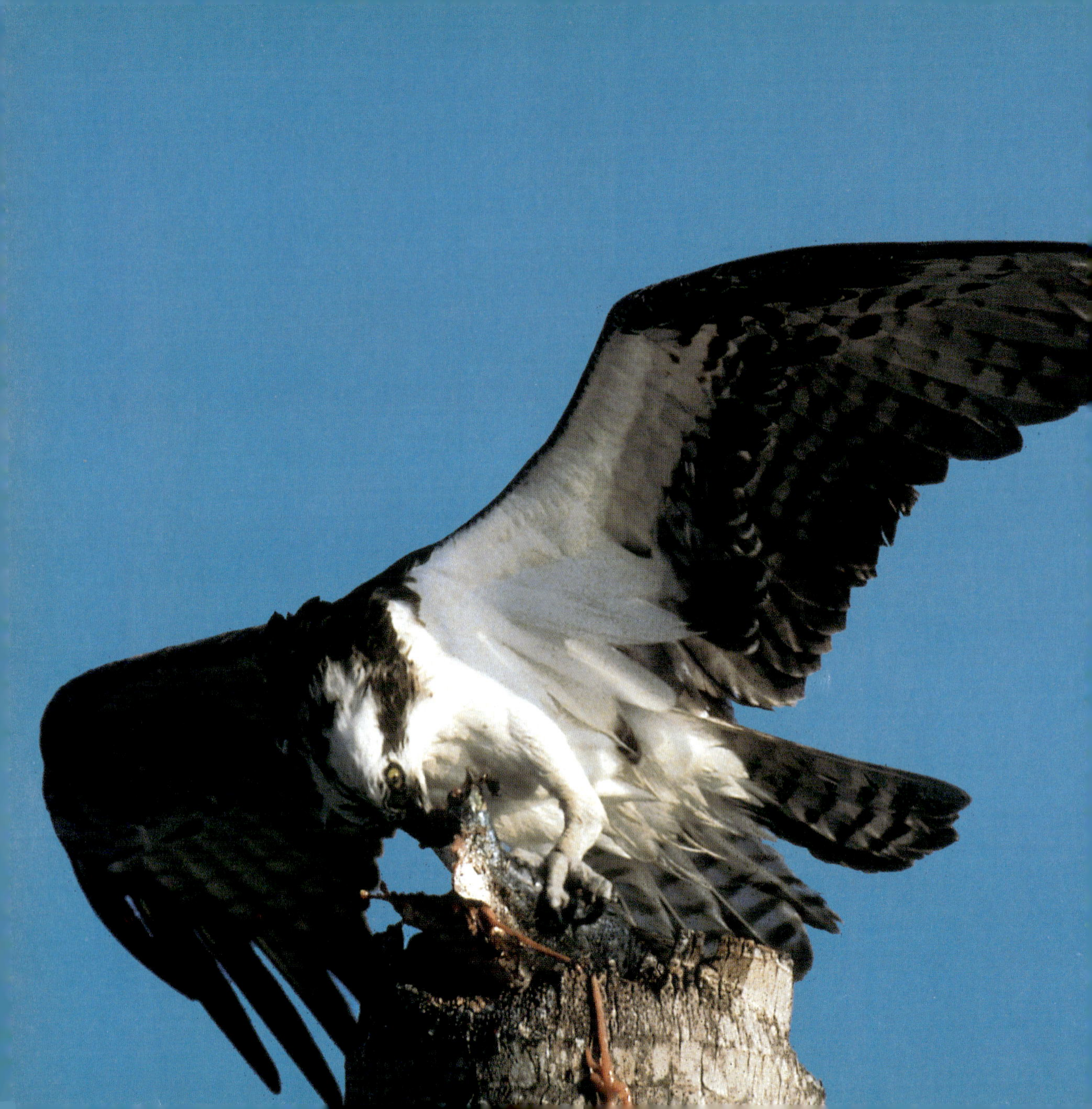

BIRDS OF PREY

Talons are so sharp and strong that they easily pierce flesh. The fistlike grip of talons kills most prey quickly.

Birds of prey also have sharp, hooked beaks. The beaks are used mostly to tear prey into small pieces.

Birds of prey are not normally a threat to people. Some of these birds, however, can be dangerous when they are near their nest.

The talons of an angry eagle or owl can puncture eyes and cause serious cuts.

While talons lock onto the catch and hold it, the osprey's hooked beak tears pieces of meat away

TOOTHLESS JAWS

The killing power of jaws with teeth is well known. Sharks, lions, and many other large animals catch prey in their toothy jaws. Some small animals have powerful, lethal jaws without teeth.

Many insects and other small, boneless creatures have mouthparts that bite and crush prey.

Two large, colorful insects with strong mouthparts are the mantises and dragonflies. In the world of insects, these predators are as fearsome as sharks and lions in their worlds.

Fearsome mouthparts of mantis munch on insect prey

TURTLE JAWS

Turtle jaws are hard, sharp-edged, and powerful. Turtles aren't fast, so they often eat dead animals and plant material. When they do kill live animals, they use their jaws to grab, crush, and tear.

The biggest turtle jaws belong to the sea turtles. The best-known jaws belong to the snapping turtles.

Snapping turtles sometimes catch ducklings and other young water birds. The snapping turtle silently swims up from below the birds and grabs its feet.

Glossary

egrets (EE grits) — wading birds with long, sharp bills, long legs, and long, delicate feathers called plumes

lethal (LEE thul) — very dangerous; deadly

talons (TAL unz) — the toe claws of birds of prey

predators (PRED uh turz) — animals that hunt other animals for food

prey (PRAY) — an animal that is hunted by another for food

INDEX

anhinga   9
beaks   5, 6, 9, 11, 14, 19
birds   5, 6, 22
   of prey   16, 19
cormorant   6
cranes   6
dragonflies   20
eagle   19
egrets   11
   reddish   11
fish   6, 9, 11, 14
herons   11
ibis   6

insects   5, 20
jaws   5, 20, 22
mantises   20
mouthparts   20
owl   19
pelicans   14
predators   5
prey   6, 16, 19, 20
talons   5
turtles   5, 22
   sea   22
   snapping   22
weapons   5